622.342 Bishop, Tom
BISHOP   Gold!

DATE DUE

JE 18	JUL 14 2010		
MR 25 '95	AUG 3 0 2011		
JA 31 '96	JUL 27 2013		
JAN 03 1997	FEB 10 2014		
AUG 1 8 2000	SEP 1 3 2014		
JUL 4 2001			
OCT 2 3 2003			
JUN 0 7 2005			
MAR 0 7 2006			
MAR 0 2 2007			

Gilpin County Public Library
Post Office Box 551
Black Hawk, Colorado 80422

DEMCO

Johnson Books
, Colorado

39204000005510

DEDICATED

To my wife Julie, who, for this book, sat in freezing water, typed, edited, typed, edited, and when finished smilingly said, "There's your book," I owe a great deal which can't really be expressed in such a short space.

Copyright 1971 By Tom Bishop

Eighth Printing 1984

Printed By
JOHNSON PUBLISHING CO.
1880 South 57th Court
BOULDER, COLORADO 80301

622.342
BISHOP

Table of Contents

I.	Introduction	Page 5
II.	Gold Panning	Page 7
III.	Do It Yourself	Page 14
	a. Rocker	Page 14
	b. Sluicing	Page 15
	c. Long Tom	Page 16
	d. Riffles	Page 17
	e. Booming	Page 17
	f. Dry Washer	Page 17
IV.	Machines Designed for You	Page 21
V.	Electronic Prospectors	Page 27
VI.	Colorado Gold Areas	Page 32
VII.	Gold Areas of the United States	Page 37
VIII.	Staking Your Claim	Page 46
	Glossary	Page 50
	Bibliography	Page 51

3

Acknowledgements

I am deeply indebted to many people who contributed their time and effort on my behalf as I plodded my way along, gathering material for this book. It would take a separate book to list all those who in some way contributed, but especially I should like to thank the following:

    Mr. Homer Stewart, U.S. Bureau of Mines
    Mr. Samuel Shepard, U.S. Bureau of Mines
    Mr. Joseph Smith, U.S. Bureau of Mines
    Dr. C.O. Frush, Colorado School of Mines
    Walt and Cathy Henderson, Idaho Springs, Colorado
    Mr. William Pyle, who as a geologist is unsurpassed, but who as a wilderness guide should be looked upon with suspicion.

CHAPTER 1

INTRODUCTION TO GOLD...

This handbook could be a gold mine for you! There's enough information and helpful hints on how to work a small gold placer in here to turn your hours of leisure into profit-making ones. In this book no attempt has been made to write "the last word" in gold mining. This book deals only with small operations that can be handled by either one or two men. The major methods are covered in this book—from gold panning to the $5,000 investment. You can learn what to look for, where to look for it, how to recover it, and what to do with it. You can learn where to purchase the equipment, where to locate it, how to operate it, and even how to build some of it yourself. After you have read this book, and if you are interested in becoming an expert, then turn to the more technical and detailed works, such as those included in the bibliography.

"Gold is where you find it." Of course it is. However, it might help if you knew a little bit about where to find it. Searching for gold in the middle of the Kansas prairie would be a hot and sweaty job. You wouldn't find much gold, either. Looking for gold in the Rocky Mountains would be a much better prospect. And we can limit the Rocky Mountain search to particular areas, also. Later in the book you will find a list of suggested areas in which to look for gold and a description of certain of the areas. But first, here's a little information that might help you.

Ore deposits are the primary sources of gold. These ore deposits are classified on the basis of their origin and structure. Other factors enter into the descriptions of ore deposits, but for a start, origin and structure are the most important for the prospector.

The more common deposits are associated with igneous activity. In other words, where one finds granite, diorite, or gabbro, plus other igneous rocks, such as dikes and sills, one will generally find gold, if the structure is right.

Structures are generally peculiar to each region in certain ways, but there is a definite relationship between gold ore deposits and intrusive rocks. Igneous dikes penetrating metamorphic rocks are sources of ore deposits. Igneous rocks are associated with mountains.

Ore deposits can be generally classified into four types:

1. Deposits that occur near fractures
   a. Quartz veins with sulfides
   b. Veins with more carbonates, silicates, or sulfides than quartz
2. Replacements
   a. Siliceous
   b. Sulfides
3. Disseminations
   a. Porphyry (coppers)
4. Placers
   a. Gossans
   b. Alluvials
   c. Sedimentary concentrations

This handbook is primarily a description of placer mining. The type of deposit described in this handbook is alluvial, which is the easiest to find and the easiest and least expensive to work.

Alluvials are found in stream beds, beaches, and fans. The types of gold found are nuggets, flakes, or dust.

The gold that occurs in the placer deposits has been carried to the alluvial, or placer, by erosive forces. In other words, the original vein, or lode, has been eroded by nature, and the gold has been carried away and redeposited.

The deposition of the gold does not necessarily mean that the vein is near at hand. Rounded nuggets have been carried far from the original source, and flakes of gold have been found hundreds of miles from their source. The presence of jagged nuggets usually means that the vein is near at hand.

With this bit of background, let's get down to business, and remember—if the fish aren't biting, and you're tired of sitting around your camp, here's another outdoor activity that could pay dividends.

CHAPTER 2

GOLD PANNING...

For the beginner, gold panning can be a frustrating experience. By gold panning, I mean working by yourself at your location, not at one of the roadside tourist attractions where the gold is planted for the tourist to "find." Buying a pan, dipping it into the stream, and sloshing it around does not take a great deal of know-how. But caring for your pan, knowing where to dip it, and how to slosh it correctly are the points with which to be familiar. There is a great deal of difference between cold, wet hands and an empty pan; and cold, wet hands and a pan trailing flakes of gold.

There are pieces of equipment that you should have with you in the field. First, of course, is the pan, which can be bought in almost any sport or curio shop. You should have a shovel, a geologist's pick, a magnet, some mercury, and a pair of tweezers. You will also need nitric acid and a porcelain bowl. These last two items will be discussed later.

Care of your pan is essential for success. The first thing that you should do when you buy your pan is to burn out any grease and oil. The reason for this is that even a thumb's worth of oil can float gold dust right over the edge and out of your pan. You can build a fire and burn the inside of the pan, or even setting it on a hot stove will generally take care of the grease. After using your pan, dry it well or it will rust. If rusting occurs, use either steel wool or light sandpaper to take out the rust; then, burn out the pan. At the end of your season, or if you are not going to be panning for awhile, smooth a thin coat of oil on it. Oiling will keep the rust off.

There are two points that you should keep in mind while you are working. Since it takes so little grease or oil to float your gold off, never put your thumbs inside the pan. If you wish to reach into your pan, don't just jab your finger into the sand and water; use your tweezers for searching and picking up.

Where does one look for gold? Unless you are a geologist, or at least very familiar with the area in which you are going to pan, you probably have a little bit of work ahead of you.

There are certain road signs that will make your work much easier. Stick to the gold mining areas. At least, you can be sure that in an area where mining has occurred, there is still gold to be found. For example, in Gilpin County, Colorado, around Central City, Blackhawk, and Nevadaville, it is almost hard not to find at least a tiny flake of gold in every shovelful of dirt. So stick to a gold mining area.

Of course, for gold panning one needs water. The smaller creeks are generally the best places for panning. More erosion takes place in a small creek that is susceptible to swollen rain water, than does in a larger creek or river. Look for the small stream that is continually buffeted by nature. It is here that gold travels more freely.

Once you have located a small stream that fits your requirements, find a miniature waterfall about ten to twelve inches high. Gold, being a very heavy mineral, sinks and works its way deep into the gravel underneath a drop-off of this nature. The perpetual force of the boiling water helps the gold to sink quickly. So in order to reach the gold, you must dig down to it. The closer than you can get to bedrock, the better your chances are to pick up something good.

Old mine dumps are excellent prospects for gold. The only problem here is that water might not be close at hand; however, if your dirt is chosen carefully, a bagful can yield a prize. Find an area on the trailing where rain has eroded the earth into a miniature canyon. Look for a place that would be a small waterfall during a rainstorm. Dig and sack the dirt and carry it to your stream where you can pan it out.

Many small, gold-bearing streams have been dredged out, not once, but several times. Of course, each spring new gold is deposited in the dredged areas, but a better bet yet would be to find an area that has not been dredged. Perhaps a gap of ten yards will separate two different mining operations. Try the gap. You might also try a shovelful of dirt from the side of the hill next to the dredging.

Panning, itself, is not the easiest thing in the world to learn to do. It takes practice just like everything else. If you can find an experienced instructor, study with him awhile. I am acquainted with a man who makes his living panning gold. It took him about three years' work to become really proficient, so don't be in a hurry. When you have filled your pan with a shovelful of gravel, dip the pan into the water and allow the water to soak in. A few shakes of the pan will send any nuggets to the bottom of it. Once you have done this, feel free to take a piece of wood and scrape off an inch of the gravel on top. Fill your pan with water again. Repeat the shaking of the pan, and then scrape off some more of the top gravel. When you have finally reached the very fine sand, use your tweezers to pick out the unnecessary pebbles in the pan.

Now, unless you discover a nugget in your pan, your work begins. You must find the flakes of gold that are in the conglomeration of sand that remains in your pan. Keep water in your pan, but not as much as before. Use a slow, circular, tilting method and allow just the lighter weight sand to slip over the side. Once you are down to just a little sand, you might find an abundance of black sand. Generally, this will be either magnetite or a mica schist. Use your magnet to pick up most of the magnetite. If you have a great deal of mica schist in your pan, your chances of finding gold are slim.

If the sand is magnetic, it is iron, which is very heavy but not as heavy as gold. You can rotate your pan and form a tail pattern with the iron. This is called "tailing." If tailing does not produce the gold, take your tweezers and poke around a little. My wife Julie did this with her first pan of sand and found her first gold flake. Not bad for a beginner!

If there is a small amount of gold in your pan, pour some mercury into the pan. The mercury will roll around and pick up the gold without picking up anything else. Then, return the mercury to its container, and start again with some new dirt.

Mercury is easiest to obtain in a thermometer. A couple of thermometers full is enough. Keep your mercury in a tightly-covered glass container.

When the ball of mercury begins to roll slowly, it is full of gold. Then you must "milk" the mercury. This is done by pouring the mercury into either a thin chamois, or a piece of heavy material. Merely squeeze the mercury through the material back into its container. Your gold will remain in the material. To separate the gold from the material is easy. Just cut away from the parent piece the material that contains the gold. Place it in a porcelain dish and pour in a little nitric acid. The acid will eat away the material and leave the gold. Follow the procedure out-of-doors and in a little breeze. The acid fumes are dangerous! Do not inhale them.

There are some other points to keep in mind. If you are not familiar with what gold looks like, visit your jewelers and examine some. More than one beginner has thought that his mica or pyrite was gold. Until you get used to seeing gold, you will probably throw some away. Part of learning to pan is learning to recognize gold. Also, if one pan of gravel should take you an hour to work through, let it. An expert can go through a pan of gravel in five minutes; but he knows what he's looking for. Let him. You take an hour.

Another important point is that of money. There are two values to gold: one, the "mint" price; two, the "specimen" price. I know a man who sold a nugget for its mint price of $100. Later he discovered that he could have sold it for $300. Today, with so little gold available on the market, almost any piece of gold can be sold as a specimen. You can contact rock collectors and jewelers and sell your gold as specimen for $95 or more an ounce. This is pretty good when the United States Mint will give you only $35 an ounce. So, take your time, and be careful.

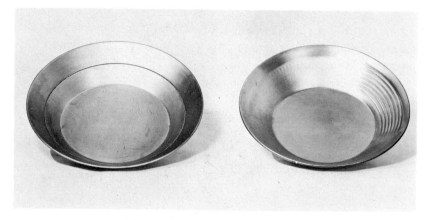

Two varieties of gold pans used.

Denver Public Library Western Collection

Sample of gold taken by gold panning.

10

Likely looking spot for gold panning.

Proper use of tweezers for gold panning.

Do not put thumbs in pan while using.

Proper place for hands during panning.

Using a magnet to take out magnetic particles.

Proper use of magnet.

CHAPTER 3

DO IT YOURSELF...

ROCKER

Once you have established a content of gold by use of the pan, you might wish to enlarge your operation somewhat. The rocker, or cradle as it is sometimes called, is the next method of least expense and simplest operation in gold recovery. It is much harder than gold panning, but it does have its advantages. Depending upon the size of your rocker and the amount of water handy, rocking could be six times more productive than panning.

A rocker is not difficult to build. One needs a hammer, a saw, a few nails, and some good pine lumber. By good, I mean make sure that your lumber has no knots or cracks. The rocker consists of only four parts: a box or trough, with two rockers set crosswise on the bottom; a screen box that receives the gravel on top of the main box; an apron underneath to catch the larger pieces of gold; and a set of riffles on the bottom of the trough to catch the fine gold.

The same holds true for using a rocker, as in the use of a gold pan. If you could find an experienced hand with the use of a rocker, you would be far ahead if you could watch him in action. But if you can't find anyone to instruct you, you can follow these general suggestions. As you go along you will catch on.

It is best to set your rocker near a source of water. Fill your screen box about three-fourths full of the gravel that you have picked for working. With a cup or dipper, pour water over the gravel; then rock your machine vigorously. Rocking will cause small material to pass through the screen and drop down onto the apron. You will have to be adding water constantly and rocking at the same time. At the end of each "rock," jerk or stop your motion. This will wash off the fine gravel in your machine. When you have run out of fine gravel in your screen box, examine the contents for larger pieces of gold. Then lift out the box and dispose of the worthless materials left and start again. This procedure can be repeated many times before it is necessary to clean out your apron and pan its contents for the fine gold.

You should also clean your riffles occasionally. This is done by running clean water into your machine and by rocking it back and forth. The fine sand will run off, and then you can scrape the remains into a pan and wash out the fine gold.

The amount of water and gravel, the slope of the rocker, and the proper rocking motion are things that you will have to work out for yourself. Those things depend upon the type of gravel in the area that you are working. You probably will have to develop new techniques for each new area that you work.

A rocker is generally used when there is an abundance of clay in your gravel. If the clay is heavy, your rocking must be even more vigorous and the amount of water used will have to be increased. Much gold will be lost if you do not free it from the clay. For a couple of days, you can expect a sore arm and shoulder from the rocking.

Now, how do you build a rocker? Rockers come in all sizes. One that is too short will lose your fine gold, and one that is too high will cause your arms a little more soreness, since you will have to heft that shovel and gravel that much more. Generally, a rocker is about 4 to 5 feet long, about 2 feet high, and about a foot-and-a-half in width. (See diagram on page 18).

SLUICING

A sluice is an apparatus that you can utilize, once you have ascertained that you are definitely in a good deposit. It does entail a little initial expense; but then again, if you really have some good gravel to work, you will come out far ahead. The work will probably take more than one man, particularly if it is necessary to have more than one box in operation. One can work much more gravel per day with a sluice than he can with a rocker, and the labor is much easier on your muscles.

There are some technical points that you must take into consideration when thinking of operating a sluice. Water is the most important facet of sluicing. You must have a continual source of water. If you don't—forget it! Water is needed to do the work, and the amount of water that you will need is important. If you were to run the average sluice for a 24-hour period, you would need at least 5 gallons of water a minute in order to successfully run every cubic yard of gravel. In other words, if you wash 10 cubic yards of gravel every 24 hours, you will need 72,000 gallons of water. Now, one cubic foot of water is worth 7.5 gallons. In order to figure out how many cubic feet of water you have to work with, you can toss a piece of wood in the creek and measure the distance that it will travel in a minute. Then measure the area of water—length times width. Now multiply the area measured by the number of feet your piece of wood covered. Then divide that by ¾. Then multiply by 7.5, and you'll know how many gallons of water pass by in a minute. Good luck!

In order not to get things fouled up along the line, you have to direct the speed of your water as it runs down your sluice. For the average sluice box of 12 feet you should have about a six inch slope. If your gravel is clay, then you should increase the slope a little so that your gravel can be broken up. A drop or two between boxes would help break up this gravel; but if you do have two or three boxes in a row, be sure that they have the same slope.

Your type of gravel will determine the slope and amount of drop between the boxes. Your type of gold will determine the number of boxes. If you have just fine gold to work with, you will need a line of 5 or 6 boxes. If you are working with nuggets, 3 boxes will be enough.

In order to eliminate such problems as large rocks and clogging of the riffles, it is a good idea to build a "lead-in" to your sluice. This lead-in is itself a sluice but is wider and deeper than your sluice, and it tapers forward to fit into your sluice. You can dump or shovel your gravel into the lead-in and pick out most of the larger rocks. This also allows your gravel to break up and flow more freely down your sluice.

The ideal method of getting rid of waste material is to wash it directly back into the stream. If this is not the case, you are in for a little work, because the waste collects rapidly. To save yourself trouble, be sure you have made some provision for removing the waste that will collect at the bottom of the sluice. You occasionally have to scrape the riffles or your fine gold will float out. When your sluice has worn out—6 months—burn it and pan the ashes for the gold. (See diagrams on page 19)

LONG TOM

A Long Tom is actually a sluice with a flume. The advantage of building a Long Tom is that the Tom, or flume, acts as a separator for the fine and coarse materials. The dirt is shoveled into the flume. The fine material is filtered through a screen at the bottom of the flume which should be about 6 to 12 feet long. The screen openings should be ¼ inch. The fine material sweeps on down the sluice and is recovered in the same manner as is done in conventional sluicing. The coarse material remains behind the screen and can be forked out. (See diagram on page 20).

RIFFLES

There are many different types of riffles that can be constructed. For instance, pole riffles are useful in hand sluicing because they are wedged in and can be removed for cleaning. Then for fine material, smaller riffles, or even carpet, should be used. The type of riffle depends upon the nature of your sluice and the gravel that you are using. To be sure, check with a local expert.

BOOMING

Booming is a method of water conservation that is employed when there is a lack of water. The water is dammed up and freed into the track as it is needed.

DRY WASHER

There are methods of recovering gold in arid or non-water areas. This is called dry-washing. The idea behind it is to feed your gravel into a pan or screen and by using a bellows, dry-wash all the useless material out of the pan or screen, and then recover the heavier gold that has sunk to the bottom.

Denver Public Library Western Collection
A 1930 WPA mining class is shown learning to operate a rocker.

Denver Public Library Western Collection
Eugene Davis, gold prospector, operating a papago (dry washer).

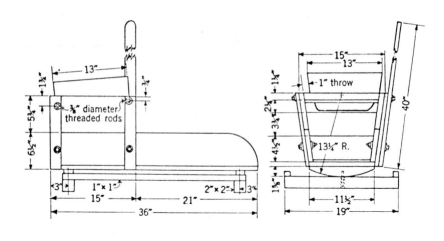

ELEVATIONS
(apron removed)

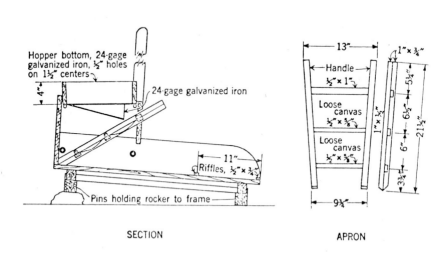

SECTION   APRON

Prospector's rocker.

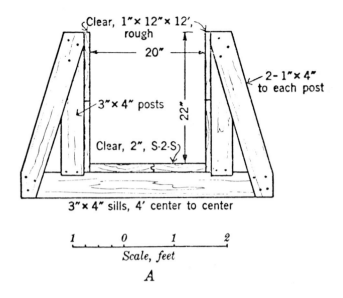

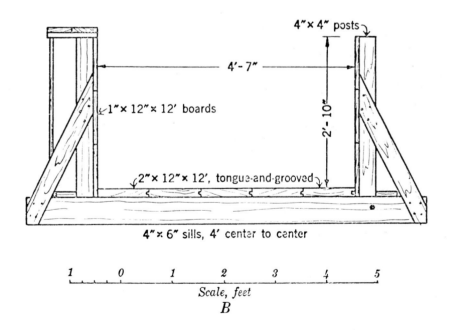

Sluice-box construction: A, Twenty-inch box; B, five-foot sluice box.

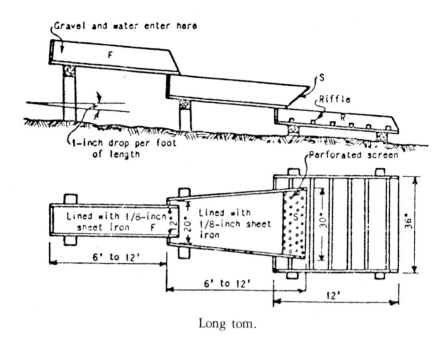

Long tom.

Denver Public Library Western Collection

Sluicing for gold along a stream.

CHAPTER 4

MACHINES DESIGNED FOR YOU...

What happens if you should make that strike of a lifetime? What happens after you have found some color, staked a claim, and had the claim assayed, and you find that you are in a deposit of gold that will really pay off? There are several mechanical gold recovery machines available on the market, and there will be one that suits your operation and your pocketbook.

Jewel Prospector

For instance, the El Dorado Prospector's Supply, of Placerville, California, carries a complete line of gold recovery machines that will fit your needs. The Jewel Prospector is a little, recovery machine that is hand operated. It weighs as little as 25 pounds but can run 500 pounds of gravel through in a single day.

Model 'A' Jet and Mojave Jet

A little bit up the line in performance, and naturally in cost, is the Model "A" Jet. This machine is power driven. It weighs 110 pounds and runs 2,000 pounds of gravel a day. (See picture on page 25.)

El Dorado also puts out a similar machine, the Mojave Jet, that will run through 2,000 pounds of gravel in an hour.

Clint Sydewinder

El Dorado manufactures other items of interest to the prospector. Classifiers, concentrators, and mills are available. The Clint Sydewinder is basically a hand-operated, mechanical, mineral concentrator. It uses the forces of gravity, water, and concentric agitation on the materials fed to the machine in a continuous operation. Although it carries the involved technical description of "Adjustable Roto-Concentric Hele-Cone Mineral Concentrator," it could be more simply called a 'mechanical panning machine.'

The heart of the Clint Sydewinder consists of a pivoted, rotating cone, capable of being tilted and locked in exactly the desired angle. Diagonally placed guides direct the materials being concentrated across the face of the cone as a spray of water is directed into the upper part of the cone.

During operation, the rotating action of the cone directs the materials in such a way that the lighter materials are washed over the lip of the cone while the heavier "concentrates" are fed to the center, where they tend to remain. When enough concentrates have accumulated, it is a simple operation to discharge the contents of the cone into your waiting container. In no time you are back in business again.

The tilt of the concentrating cone and the speed of rotation, coupled with a balancing flow of water, make for a versatile, yet simple operating machine. Recovery of concentrates in the 200-minus class right along with much coarser materials is common practice with the Clint Sydewinder.

Although the Clint Sydewinder weighs only about 40 pounds, its low profile and construction features make it considered to be quite stable in its various attitudes of operation.

Whirlovac Vacuum Impact Mill

El Dorado's Whirlovac Vacuum Impact Mill can be of great value to you. By running hundreds of samples, the manufacturers have found fine gold and other metals in practically every form of rock known, including high grade silica and bull quartz often in paying quantities. In order to release this fine gold and other metals without grinding or sliming, it becomes necessary to impact the ore by a dry process from 100 minus 300 mesh or more. All of the rock is shattered, leaving the metals free. By this method all of the gold is rolled or balled into small nuggets, leaving it in its natural state. They have a special nugget trap built into this vacuum impact mill that catches all the coarse gold and other metals.

This mill has a built-in circuit that does not clog, thus allowing the ore to whirl around inside, and as it becomes finer and finer it is carried higher and higher in the circuit.

In the top of the duster there is a special variable vacuum tube that can be adjusted from a light to a strong vacuum. By this method the size of the grains of material varies from $\frac{1}{8}$ up to 500 mesh or more and can save the fine gold and other metals.

Damp or wet rock can be run through this machine then run directly over an "Air Float Concentrator" or a wet concentrator by adding their "Drier," which is very easily and quickly attached, also which is very inexpensive and easy to operate. (See picture on page 25.)

Denver Mechanical Gold Pan

The Denver Equipment Company of Denver, Colorado, is a leading manufacturer of gold recovery machines. The Denver Mechanical Gold Pan recovers coarse and fine values amenable to concentration by gravity, matting, or amalgamation. The oscillating motion duplicates "hand pan" action. The capacity ranges up to 2 cubic yards per hour and a nugget trap is provided. A wash spray in the feed hopper cleans and disintegrates the feed. It is of a rugged, compact, and portable design. The unit is easily transported in a trailer or pickup truck, and is assembled on heavy skids for easy relocation. This is a complete, self-contained placer plant. The low initial cost for this complete machine, plus minimum operating cost and maintenance, combine to make this a most economical placer unit. The number 24 Simplex, is driven by a three horse power, gasoline engine. (See picture on page 26.)

Denver Portable Suction Dredge

The Denver Portable Suction Dredge unit with your Denver Mechanical Gold Pan will enable you to operate at full capacity by hitting "pay dirt" beyond the reach of manual handling. This unit consists of a gasoline-engine driven, 2 inch sand and gravel pump mounted in a water-tight box with carrying handles and equipped with a 20 foot suction hose with an adjustable probe. It was designed for maximum portability and can easily be carried by two men or packed on mule back. Its rugged construction affords full pumping efficiency under the roughest of conditions. An adjustable end plate on the pump permits compensation for wear, and replacement of parts is easily accomplished. With the pump turning at 1800 r.p.m., the dredge will handle between five and six yards of straight sand per hour. The suction pump is capable of raising gold or other heavy minerals from depths of 16 to 18 feet. This depth ability, capacity, and elimination of manual labor enables the unit to pay for itself in a short time.

Denver Trommel Jig Placer Unit

The material to be treated with this technique is shoveled or dumped into the large feed hopper and then is fed regularly into the washing and disintegration section of the rotating trommel. Spiral lifting blades elevate and mix the gravel several times during a revolution. Water is added at the feed end of the machine and the action secured by lifting blades breaks up the lumps of cemented or clayey materials so that the desired minerals are freed and separated. The material then passes to the screened section for the removal of the washed large particles from the undersize. This section includes an inner screen made of heavy steel plate having rectangular openings. Surrounding this heavy perforated plate is an outer replaceable steel screen of 3, 4, or 6 mesh. The oversize material is discarded at the end. The unit is driven by a gasoline engine or electric motor which also drives a centrifugal water pump, furnishing water for washing in the trommel as well as for use in the jig operation. A steel wire reinforced, rubber suction hose and a foot valve are furnished. Mounted on a steel base with steel sides, the unit is easily transported from place to place and the trommel and jig are supported on a structural steel frame which carries necessary shafting and belting to make the unit complete and self-contained. The No. 2 Trommel has a capacity of 1-2 cubic yards per hour. (See below.)

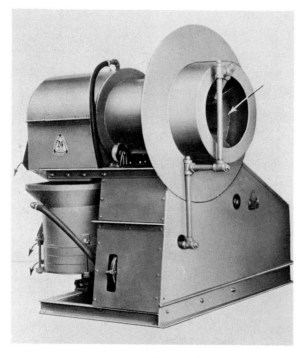

Denver Trommel

Model A "Portable Jet"

Whirlovac

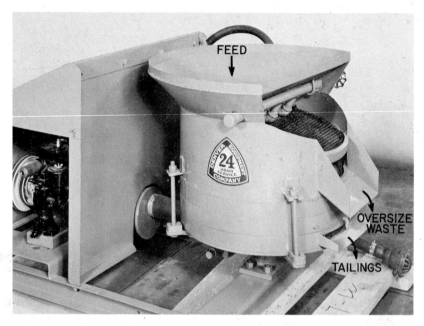

Denver Mechanical Gold Pan

Picture showing a Denver Mechanical Gold Pan in operation.

CHAPTER 5

ELECTRONIC PROSPECTORS...

The prospector and his burro of 100 years ago did a pretty thorough job of covering a lot of country and finding a lot of mineral. Today's prospector doesn't have the time to wander through the canyons for weeks or months, examining every rock and crevice. He is only a weekend prospector, but he has available to him devices that aid him in his quest for riches.

Mineral detectors are electrical metal finders that the modern prospector can carry with him. They will do the work of finding the mineral that the old prospector had to find by himself.

Coinmaster

White's Electronics of Sweet Home, Oregon, sells a complete line of detectors. The Coinmaster line will detect non-ferrous or non-magnetic metal such as coins, jewelry, gold, silver, copper, and platinum. These models are able to detect a coin the size of a dime as well as nuggets. Larger metal objects are more readily detected. It is ideal for probing around old ghost towns, mining camps, or any other place where there might be hidden treasures. All the Coinmaster models use a gear-down vernier radio tuning control. All models are 4 ¾ inches by 10 inches by 4 inches. The Coinmaster Model No. 1 and Model No. 2 both use genuine Brazilian quartz crystal with gold plated electrodes.

The Coinmaster Model No. 1 has a four transistor chassis and is powered by one Eveready No. 246 9-volt battery. Weight: 3 lbs. 4 ozs.

The Coinmaster Model No. 2 includes a built-in speaker, a battery-check meter, and is powered by two Eveready No. 246 9-volt batteries. Weight: 4 lbs. 2 ozs.

The Coinmaster Model No. 3 is designed for dependable year-round use in the more extreme climates. The depth detecting ability of sensitive instruments can vary with drastic temperature changes; however, this model has built-in, thermal-insulated condensers, giving greater stability and higher sensitivity under all weather conditions. It is a five-transistor instrument that has an encased radio tuning condenser, a silver mica condenser, and a built-in speaker. There is a combination battery check and an indicating meter which registers the intensity of the signal strength when detecting a metal object. It is powered by two M6 9-volt batteries. Weight: 4 lbs. 8 ozs. (See picture on page 31.)

The Coinmaster Model No. 4 is designed to detect also at greater depths on larger objects. This model uses an all-transistor, solid state circuitry. It has a battery check. A "Mineral-Null-Metal" control enables the operator to detect both non-magnetic and magnetic objects. When the instrument is operating in the mineral position, magnetic objects will be detected such as mineral float, magnetic veins, and ore deposits containing magnetite which so many gold deposits contain. Incorporated is a radio tuner and a volume control. Weight: 4 lbs., 6 ozs.

Goldmaster

White's Electronics also produces the Goldmaster line of detectors. These detectors are designed more specifically for the mineral prospector.

The Goldmaster Model 66-T will detect a natural gold nugget as small as a grain of corn through quartz rock. This instrument will detect both small and large objects. It comes with two loops: The 7 ½ inch loop is designed to get into smaller places and to locate small single nuggets. The 11 inch loop is for detecting larger objects at more depth, and is used for vein tracing. It will detect through ice, snow, dirt, wood, mud, glass, water, concrete, some stone asphalt, plaster, adobe, brick, and other non-metallic objects. The 66-T is sensitive to both magnetic and non-magnetic metals, is an all transistor instrument with a speaker and volume control, and has a built-in battery checking meter circuit. Weight approx. 6 lbs.

There are other variations of the Goldmaster. The Goldmaster 66-T Special has one 7 ½ inch triple loop. All three loops operate at the same time. Weight: approx. 5 lbs. 4 ozs.

The Goldmaster S-63 incorporates 3 low-drain miniature electronic tubes, genuine Brazilian quartz crystal with gold plated electrodes, and a special 4-transistor low-drain amplifier. It is designed for checking outcroppings. It has a built-in battery tester, and comes complete with a 3 ½ inch loop for nuggets and a 12 inch loop for general prospecting. Weight: 6 lbs. 2 ozs.

The Goldmaster S-64 is the same as the S-63 except that in the S-64 a 2½ inch signal intensity meter is used in addition to the built-in speaker. Weight: 6 lbs. 4 ozs.

In addition to the features of S-64, the Goldmaster Model S-65 has a protective grill over the speaker, a 4½ inch signal intensity meter in place of the 2 ½ inch meter, and three loops instead of two. Weight: 6 lbs. 8 ozs.

Nuggetmaster

For the prospector with just a little more money to spend is White's Nuggetmaster.

The Model TR Nuggetmaster is all transistorized and incorporates large Dual 4 ½ laboratory type indicating meters. One meter is engineered to read on soft metals, which include gold, silver, copper, coins, and other conductive metal objects. The other meter is designed to read any detectable magnetic field, as so many mineral deposits and pieces of float are. Two all-weather loops come with the instrument—a 7½ inch loop and an 11 inch loop. It features a Mineral-Metal Selector Switch, an Automatic Selector Switch, a special Vernier Radio Tuning control, four meter ranges of sensitivity, and is powered by 2 small economical 9 volt dry batteries. Weight: approx. 6 lbs.

D-Tex Standard

D-Tex Electronics of Garland, Texas, is another producer of metal detectors. The D-Tex Standard detects through concrete, rock, wood, glass and plastics, comes equipped with headphones, 6 inch and 12 inch fiberglass search coils and with a battery installed and ready to go. The leg is two piece for compact packing or carrying in a backpack, or, the unit may be used on the lower section of the leg only to get into those close places. It will detect both ferrous and non-ferrous metals. It is also equipped with a control to tune for either metal or mineral. Weight: 2 ½ lbs.

D-Tex Delux

The Delux is similar to the Standard but with more amplification added for more power and additional range. The Delux has a speaker as well as an Indicating Meter to aid in ease of detection. It has a volume control that controls the amount of sound only. It does not cut down on range and sensitivity. A headset is also included that may be used if desired. When plugged in, it automatically turns off the speaker so only the operator knows if something is detected. Weight: 3 ½ lbs.

D-Tex Professional

The D-Tex Professional has a loud toned speaker. It also comes equipped with a headset. Plugging in the headset automatically shuts off the loudspeaker. The indicating meter has a built-in light for night use. The speaker may be turned off and you may use only the lighted meter or you may use the headset and the lighted meter. This gives completely silent operation with good visibility of the meter with no outside light. Weight: approx 5 lbs.

Goldak 820 Commander Jr.

The Goldak Company, Inc., of Glendale, California, puts out three detectors that can be used for finding gold. The Goldak 820 Commander Jr. detection head incorporates a transmitter and a receiver antennae in an inductive-balance circuit concept which allows the operator to detect small or large metal objects buried from the surface to a depth of five feet. Weight: less than 4 lbs. Model 820S has built-in speaker for audible tone.

Goldak Commander 720

The package design of the Commander incorporates a guided telescoping handle which enables children as well as adults to effectively use the instrument. Other engineering considerations along this same line include a single combination on-off switch and sensitivity control which is manipulated by the operator. Also included is a Tell-Tone speaker system which provides a tone when buried metal is detected. This same indication can be seen on the built-in meter or heard in the earphones which are provided with the instrument. Depth range is in excess of 5 feet. Effective depth range for placer gold deposits is 36 inches. Weight: less than 4 lbs. (See picture on page 31.)

Goldak Bonanza 1100

The circuits incorporated in this locater include 7 transistors, three R.F.

Signal transformers, two audio power transformers, plus numerous deposited carbon resistors, and metallized Mylar capacitors. These circuits are housed in Cycolac plastic cases. Approximate depth range for an ore vein or placer is 20 feet. Weight: 9¾ lbs.

The Frontiersman

Relco Industries of Houston, Texas, features two detectors that will aid the prospector in his journeys. The Frontiersman will detect small coins, rings, nuggets, placer gold deposits or pockets, rich veins or any other kind of metal or detectable mineral. The piece is powered by one small battery. It is fully transistorized, and it has a built-in clear tone loudspeaker. Weight: approx. 3 lbs.

Coin-Ranger Professional

This piece is easy to operate and will detect either ferrous or non-ferrous metals. It can distinguish between gold bearing magnetite, and metals such as copper, iron, gold, and silver. This can save much digging and add to your profits. A selective tuning control permits adjustment, so that it will not detect bottle caps, nails, and other small metal objects, but will sound within the detection range of placer gold deposits. Weight: approx. 2 lbs.

Coinmaster Model No. 3   Goldak "Commander" Model 720

CHAPTER 6

COLORADO GOLD AREAS...

Within the state of Colorado there are numerous areas in which one could prospect for gold. A quick check with the United States Geological Survey, or even a Colorado history book, would render the names and locations of numerous gold districts. Once you have decided upon an area for prospecting, you should read a geologic report of that area in order to familiarize yourself with its potential. This will save you a lot of wandering. You can determine from your reading the most likely spots before you drive to the location. These reports, published by the U.S.G.S., can be secured from the U.S.G.S., or found in most large libraries. Even though there is a great deal of technical language employed in the reports, the lay reader can easily pick out what is necessary. The following are gleanings from four major reports on goldmining in Colorado.

Central City, Blackhawk, Nevadaville

Gilpin County, in which the old mining towns of Central City, Blackhawk, and Nevadaville lie, is one of the most productive areas available for the panning enthusiast. If you are in or around one of the areas which has been mined extensively, you could have at least a flake of gold in every pan.

The veins in this district can be found along small faults in the rock. Some of the veins run in an east-west direction, while others have a tendency toward a northeast-southwest direction. There is an abundance of pyrite, chalcopyrite, and galena associated with the veins. Also, there is a great deal of iron in the form of magnetite within the area. This helps in the location of gold. If your black sand is magnetic, you are in a gold area. If your black sand is not magnetic, go somewhere else because you are in a mica schist which does not associate itself with gold in this region.

Because of the tremendous amount of mining activity in the area, you can readily pick a spot in any number of the gulches around these towns, and be in good position. One can pan a shovel load of dirt taken right in the town of Blackhawk and find color.

The better areas for work are downstream from the three towns. This is only because the areas toward Nevadaville and beyond do not contain any water. Otherwise, they would probably be the better since so many people have already worked below town. If one does not mind sacking dirt to a creek, the areas south of Central City are good for panning. The mine dumps of this region could also be productive.

Georgetown, Silver Plume, Idaho Springs, Empire

Georgetown, which, in a neat, little valley along the banks of Clear Creek on I-70, was once a proud producer of some gold and a lot of silver. Leavenworth Mountain which is just south of Georgetown was the major area of gold producing ores. The veins were good ones and produced a lot of high grade ore. However, most of the remaining area around Georgetown derived its riches from silver veins. The first discovery of any type of ore was gold, which George Griffith stumbled across on the banks of South Clear Creek in 1859.

Within the same area is Empire, which was at one time a good gold producing area. Empire is on highway 40 along the banks of West Clear Creek. Just east of Empire, West Clear Creek and Clear Creek join together. Below this junction of the two creeks would be your best bet for a little work.

Even though Clear Creek below the junction of the two forks is a rather large body of water for panning, one can delve into the small eddies and miniature falls and, after digging deeply, can find a trace of gold. In this particular area, a sluice would probably be more to one's liking.

Since the new highway has juggled Clear Creek around somewhat, panning below Georgetown would not be profitable—unless one wished to carry his dirt from the old creek bed to the new and wash it there.

West Clear Creek below Empire would be the better of the two spots. West Clear Creek is not so large as Clear Creek and likely places would be easier to find. West Clear Creek would also be better because of the amount of gold-bearing veins that occur in the Empire area. The gold-bearing veins which predominate in the Empire area lie to the west, rather than to the east of Empire. On Lincoln Mountain, just south of Empire, there were good gold producing veins, and the small streams running down from this area are potential beds for nuggets.

For the panner, Silver Plume can be eliminated as a likely prospect for gold. The veins are rich in silver, but have little or no gold.

Many of the primary veins in the Idaho Springs area contain large amounts of gold. Just north of Idaho Springs, between Idaho Springs and the Central City area, the veins contain gold. This area would possibly be the best for panning. Seaton Mountain is primarily a silver producing area, but elsewhere in this area the veins are gold. In fact, within the Georgetown-Idaho Springs area, this district north of Idaho Springs would be the best place for panning. Clear Creek below Idaho Springs has been dredged a little. But, again, Clear Creek is better ground for a sluice than for panning.

Breckenridge

Aside from being a fishing and skiing mecca, Breckenridge, just over Loveland Pass on highway 6 west of Denver, is also a potential gold area. The first discovery of gold in the area was made in 1859 in Georgia Gulch. This was a placer venture in which the gold was taken from an alluvial deposit.

The Breckenridge area is divided into two different mineral zones: an inner area that borders the Blue River, which is a gold area; and an outer zone of mineralization which is primarily silver in nature. The eastern part of North Star Mountain, for general purposes, represents the center of the area. The faults which occur in the mineralized areas are small, but the veins occur in them. The veins, or lode deposits, are secondary—highly weathered and eroded.

The Breckenridge area has a great deal of alluvium, terrace gravel, and glacial moraine. This, in connection with the highly weathered and eroded veins, gives the gold seeker a good chance of turning something up with his pan.

The gold seeker would do well to work in the small streams around the center area, which occurs on Monte Cristo Creek. The mine dumps in this area are also good sources for nuggets. Every spring someone stumbles across a small nugget on or around the many old mines in this area.

Another good possibility would be to locate yourself on a small creek below one of the small faults that harbors a vein. In all probability the erosive forces will have dislodged several nuggets from their source, and they will have traveled down the creek.

The Blue River and Swan Creek have been dredged thoroughly, but this started seventy years ago and one could possibly find a likely spot on either body of water that would yield a nugget.

Any of the small creeks in which there is alluvial, are good prospects.

The area is actually better suited for a rocker than for a pan. There is definite clay material in the area, and a rocker would break up the clay which surrounds the nuggets. A pan would lose several nuggets along the way.

Cripple Creek

Cripple Creek lies southeast of Hartsel, which is in South Park on highway 24. Cripple Creek was cattle country until cowboy Bob Womack stumbled across some gold in one of his "prospects." Then, cattle were replaced by miners, and Cripple Creek became one of Colorado's richest bonanzas.

Gold has been taken from both veins and placers in the Cripple Creek area. The veins are the primary and most important source of the metal. The veins occur in fissures in all of the country rock. However, they are associated with minor faults that are in this area.

The erosion in the area that produced the placers has not been great since the last intrusion of the eruptive rocks. However, the erosion has been great enough to slide much of the gold off the hills and into the gullies that circle the hills. The gold, because of its weight, has easily worked its way down into the gravels of the gullies and into other low places.

The main area of concentration for the panner should be north or west of the town itself. Many of the best placers were in this area. The low spots in the gullies and just under small drop-offs would prove productive. However, in many of the small gullies, you would have to sack dirt and carry it to water. To eliminate this, you could stick to the head of Cripple Creek, which is a good spot for panning.

Below the town, many small placers have been worked. Beaver Creek, Arequa Gulch, and Wilson Creek are areas of color. Placers were worked in these areas, but they were all given up because of the low content and lack of water.

Cripple Creek is a rich area and certainly the panner would have good luck in the area.

There are other potential areas for investigation within the state of Colorado:

Leadville-Lake County	Ouray-Ouray County
Fairplay-Park County	Creede-Mineral County
Eagle-Eagle County	Lake City-Hinsdale County
Silverton-San Juan County	Bonanza-Saguache County
Telluride-San Miguel County	Rico-Dolores County

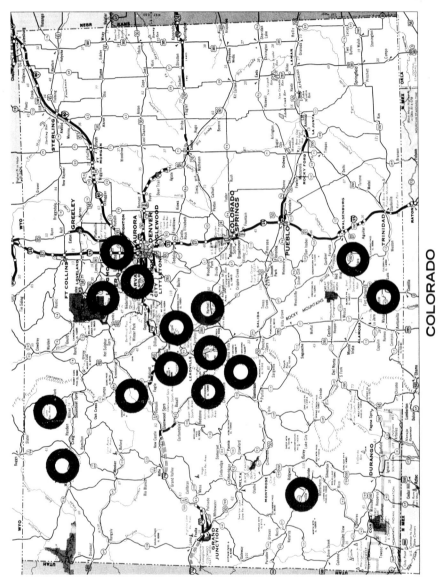

CHAPTER 7

GOLD AREAS OF THE UNITED STATES...

There are several states that have gold-bearing areas. Following are maps of these states and the significant areas within the states where a prospector might come up with a deposit of his own:

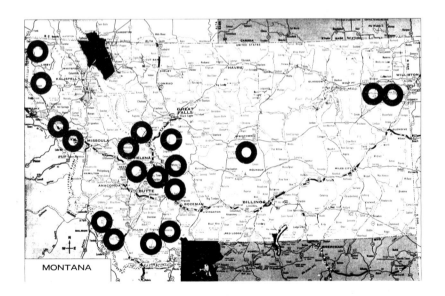

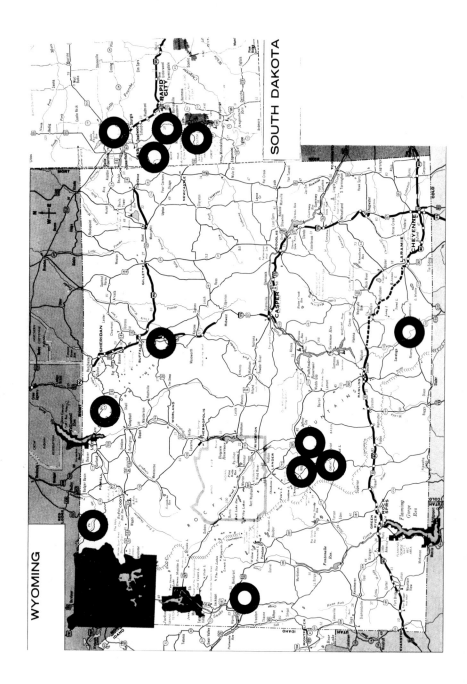

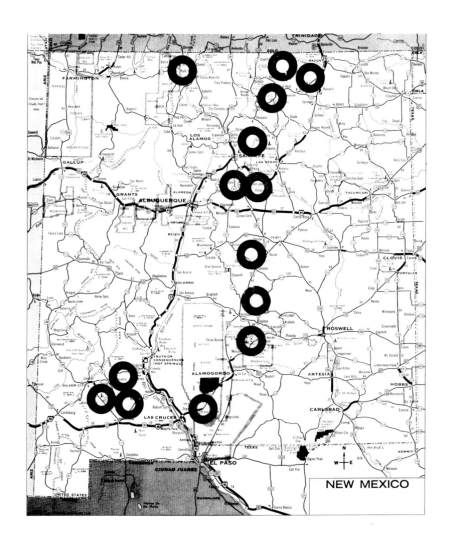

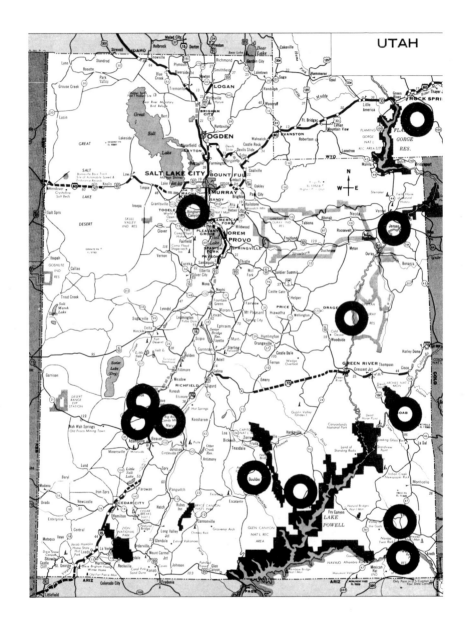

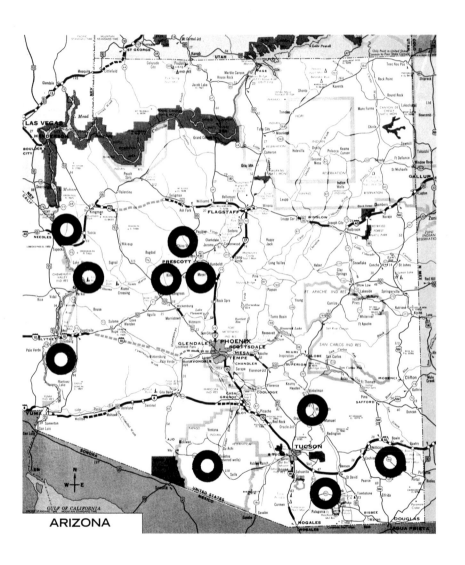

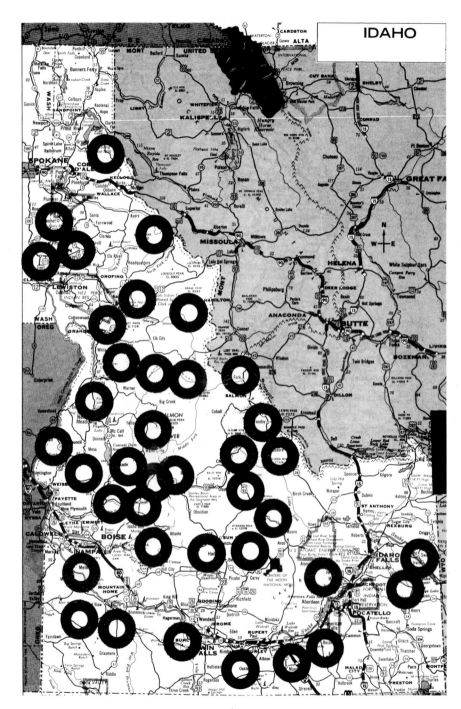

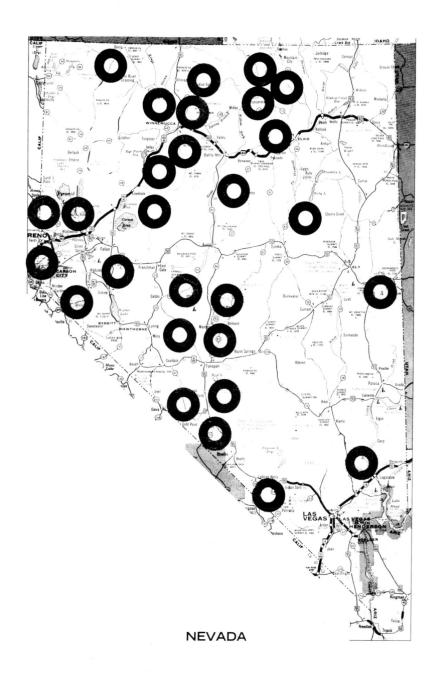

NEVADA

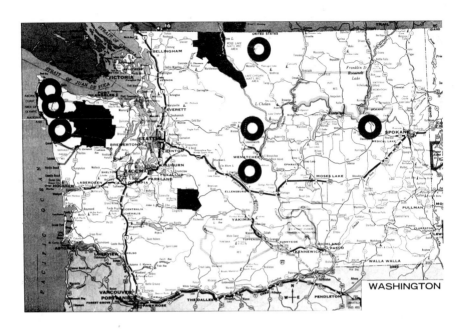

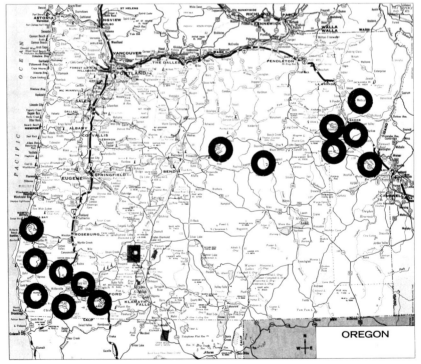

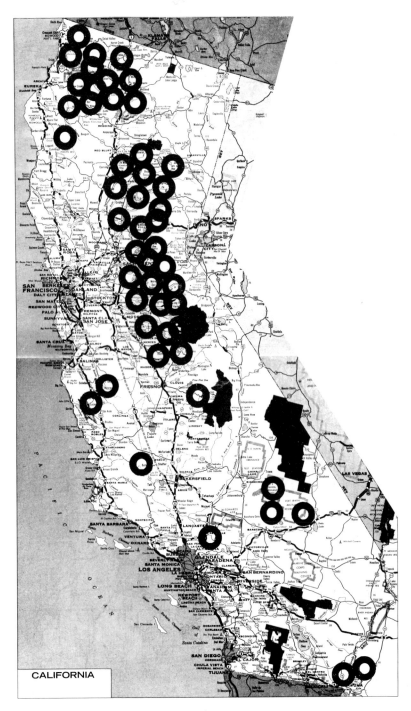

CALIFORNIA

CHAPTER 8

STAKING YOUR CLAIM...

What to do after you have made your strike? The first thing to do is to take into consideration what you have in your deposit. Deposits of coal, oil, gas, oil shale, sodium, phosphate, and potash belong to the United States and are not subject to location under the United States mining laws. Whatever is recognized as a metallic or other substance found in public lands in quantity and quality sufficient to render the land valuable, is considered legitimate under the mining laws. In other words, no sand, stone, gravel, pumice, pumicite, or cinders constitute legality under the mining laws.

The second necessary thing that you must do, is to make sure that your deposit can legally be claimed. Any vacant, public surveyed or unsurveyed lands are open to prospecting and to location and purchase. Mineral lands in forest reservations are legal areas for claims. There are special requirements for national parks and monuments. Anyone who is seriously contemplating a location should have immediate access to Circular No. 2149, printed by the U.S. Department of the Interior, Bureau of Land Management. Some of the information contained in that circular will be covered generally here, so that the reader will have a basic understanding of the laws and requirements.

Locations can be made in the following states: Alaska, Arizona, Arkansas, California, Colorado, Florida, Idaho, Louisiana, Mississippi, Montana, Nebraska, Nevada, New Mexico, North Dakota, Oregon, South Dakota, Utah, Washington, and Wyoming. Check with your state government to be sure of the requirements. In Colorado, be sure you have Circular No. 3 from the Bureau of Mines, 215 Columbine Building, 1845 Sherman Street, Denver.

Now then, who was first? Were you there first, or was someone else? Check with the county assessor to make sure. Once you have established that your location can be yours, you must within thirty days from the date of discovery, record your claim in the office of the county recorder. The thirty-day period begins when you first post a notice on your claim that it is your discovery. This notice should contain the name of the placer, your name, date of location, and the number of acres claimed. At the time of location you should mark the boundaries with posts, one at each angle of the claim.

No placer can exceed twenty (20) acres for each person. In case of more than one person, no placer can exceed one hundred and sixty (160) acres.

Your placer will have to be surveyed. The law states that placers shall conform as nearly as practicable with the United States system of public land surveys and the rectangular subdivisions of such surveys, whether the locations are upon surveyed or unsurveyed land.

Locating your claim is not impossible. In fact, it is relatively simple. First, get yourself a United States Geological Survey map of the area in which your claim is located.

For most of the state of Colorado, you will have two reference points from which to start. The intersection of these points is called an "origin." One of these is the 40th Parallel. The other is the 6th Meridian. Most Colorado land is measured and identified from the origin of these two lines. The next divisions, so to speak, are called "townships." These townships are square units of land 36 square miles in area that are measured from the origins. Township 5 South, is the 5th township south of the base line, or the 40th Parallel. Range 74 West is the 74th township west of the sixth Meridian. In other words, count five townships down from the 40th Parallel, 74 townships west of the 6th Meridian and you are in Township 5 South, Range 74 West, which is abbreviated as T5S, R74W. You are now in an area of 36 square miles.

The townships are divided into 36 areas called "sections." They are numbered in this manner:

R74W

1	2	3	4	5	6
12	11	10	9	8	7
13	14	15	16	17	18
24	23	22	21	20	19
25	26	27	28	29	30
36	35	34	33	32	31

T5S

Let us say that we are locating a ten acre piece of ground in this township. Our ground is in section 15. By looking at the map, you can easily find section 15, T5S, R74W.

R74W

1	2	3	4	5	6
12	11	10	9	8	7
13	14	15	16	17	18
24	23	22	21	20	19
25	26	27	28	29	30
36	35	34	33	32	31

T5S

You are down to 640 acres now. This section of 640 acres is divided into four squares:

Section 15 R74W, T5S

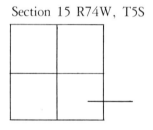

Each quarter is 160 acres in area. These quarter sections are once again divided into four parts, each part having 40 acres:

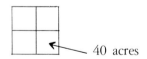
40 acres

And, of course, these quarter sections can be broken down into four equal parts, ten acres apiece:

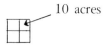

You have just located the top of Gray Wolf Mountain, elevation 13,610 feet, Clear Creek County, Colorado. Your claim can be located just as easily.

There are other items with regard to your deposit that you should keep in mind. You will have an annual one hundred dollars' worth of assessment work to do in order to hold title to your claim. This work begins on the meridian of September 1 succeeding the day of location. Any geological, geophysical, or geochemical work will be considered as "labor." This and other important information can be found in the two circulars already mentioned. Your total assessment work, if the claims are held by more than one person, can be done on one claim for all involved. In other words, if three of you held 60 acres, the $300 of work could be done on one claim.

In order to ascertain the quality of your deposit, you must have an assay of the material. If the deposit is large enough, the United States Bureau of Mines will do this for you. You could also utilize the services of Colorado Assay or Denver Assay. There are many independent assayers that are quite reliable, but sometimes they do not have the facilities available.

You have just taken yourself through a very quick course in gold recovery. Perhaps all that you need now is the actual experience. Or maybe you will wish to read more about this fascinating business.

If it's reading you want, I refer you to the bibliography. If it's experience you want, I refer you to the mountains—and good luck!

GLOSSARY OF TERMS...

alluvial deposit—sediment deposited by stream.
alluvial fan—deposit of alluvium deposited by decrease in stream's energy.
carbonate—rocks translucent in nature and not hard. Soluble in acid.
dike—intrusive igneous body cutting across layers of earlier rock.
diorite—like granite, only plagioclase, and little quartz.
gabbro—igneous rock in which dark minerals such as pyroxene exceed 50 per cent of the rock.
gossan—oxidized outcropping of limonite and gangue.
granite—igneous rock containing feldspar and quartz.
igneous—rocks whose origin pertains to a molten state.
intrusive—igneous rock that flows between older rocks.
metallic—chemical elements producing lustre, hardness, and other metallic qualities.
metamorphic—rock transformed from one type to another through heat or pressure.
non-metallic—solids, liquids, gases.
ore deposit—primary source of mineral. Occurs in a vein. Associated with igneous rocks.
placer deposit—secondary source of mineral. Relocated from primary source by physical forces.
porphyry—rock containing large grains in a fine-textured ground mass.
sedimentary—rock that has been redeposited from original source.
silicates—minerals whose chemical nature is $SiO2$ (quartz).
sill—intrusive igneous body running parallel to older layers of rock.
sulphides—heavy, brittle, compound of element with sulphur. $Ag2S$- Argentite.

BIBLIOGRAPHY...

Bateman, Alan M. *Economic Mineral Deposits.* New York: John Wiley & Sons, Inc., 1958. pp. 916.

Boericke, William F. *Prospecting and Operating Small Gold Placers.* New York: John Wiley & Sons, Inc., 1960. pp. 145.

Cross, Whitman, R.A.F. Penrose, Jr. *Geology and Mining Industries of the Cripple Creek District, Colorado.* Washington, D.C.: Government Printing Office. pp. 209.

Keating, Paul H. *How to Pan Your Own Gold.* Denver: The Geiger Center, 1953. pp. 47.

Lovering, T. S. *Geology and Ore Deposits of the Breckenridge District.* Washington, D.C.: Government Printing Office, 1934. pp. 64.

Pough, Frederick H. *A Field Guide to Rocks and Minerals.* Boston: Houghton Mifflin Company, 1955. pp. 349.

Spurr, Josiah E., George H. Garrey, Sidney H. Ball. *Economic Geology of the Georgetown Quadrangle.* Washington, D.C.: Government Printing Office, 1908. pp. 422.

Thorne, W.E., A.W. Hooke. *Mining of Alluvial Deposits.* London: Mining Publications, Ltd., Salisbury House, 1929. pp. 171.

Lode and Placer Mining Claims, Tunnel Sites and Mill Sites. Circular No. 3, Bureau of Mines, State of Colorado.

Regulations Pertaining to Mining Claims Under the General Mining Laws of 1872, Multiple Use, and Special Disposal Provisions. Circular No. 2149. U.S. Department of the Interior, Bureau of Land Management.